Cover and internal design by Will Riley
Internal images © makyzz/Freepik, vectorpocket/Freepik, macrovectr/Freepik
Sourcebooks and the colophon are registered trademarks of Sourcebooks.

Published by Sourcebooks eXplore, an imprint of Sourcebooks Kids
P.O. Box 4410, Naperville, Illinois 60567-4410
(630) 961-3900
sourcebookskids.com
First published as Red Kangaroo's Thousands Physics Whys: *A Shocking Idea: Electricity*
in 2018 in China by China Children's Press and Publication Group.
Library of Congress Cataloging-in-Publication Data is on file with the publisher.
Source of Production: PrintPlus Limited, Shenzhen, Guangdong Province, China
Date of Production: June 2020
Run Number: 5018751
Printed and bound in China.
PP 10 9 8 7 6 5 4 3 2 1

Let's Power Up!

Charging into the Science of Electric Currents with Electrical Engineering

sourcebooks eXplore

#1 Bestselling
Science Author for Kids
Chris Ferrie

"Ouch! That hurt!" Red Kangaroo cries. "Dr. Chris, can you tell me why I sometimes get shocked when I touch something made of metal?"

Dr

"That shock happens because of electricity," Dr. Chris replies. "Electricity is what causes lightning, and it turns on the lights in your home."

“I know about electricity!” says Red Kangaroo. “It also comes out of the outlet in the wall. My mom says I should never put anything inside it because that would be dangerous!”

“Your mom is right,” Dr. Chris says. “Safety first!”

“Electricity exists because of two things called protons and electrons. An **electron** has one negative charge and a **proton** has one positive charge,” says Dr. Chris.

Dr. Chris

"I know about these!" Red Kangaroo says. "Positive likes positive... No, negative likes negative... Oh, I can't remember."

“Here’s a little trick to help you remember. Think ‘opposites attract,’” Dr. Chris says. “This means that electrons and protons will always want to be near each other. But electrons will push away other electrons and protons will push away other protons.”

“So positive likes negative and negative likes positive! That’s easy!”

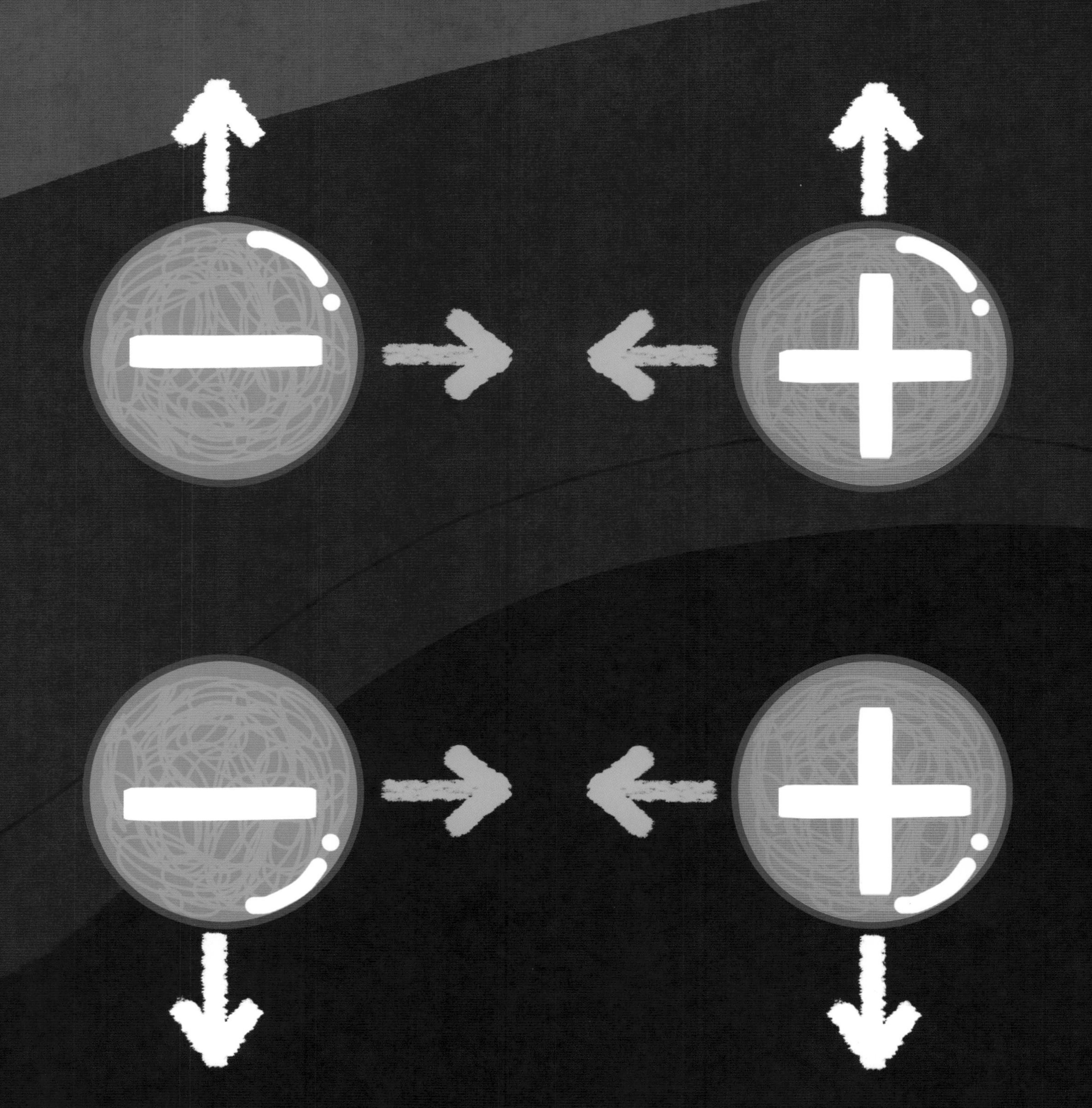

"Protons like to stay in one place. And electrons like to move around," Dr. Chris continues. "Sometimes electrons move by flowing through something."

"Like water flows through the tap or a hose?" asks Red Kangaroo.

“Yes! Just like that!” Dr. Chris replies. “The movement or flow of the charges is what makes **electricity**! Electrons flow in a loop. The outlet in the wall has two holes that are there to break the loop and stop the flow of electrons. The loop closes when you plug in an electric cord.”

“And then the light can turn on!” says Red Kangaroo. “That explains why I shouldn’t play with an outlet. But what about the **shock** I got while closing the door?”

"Don't worry, I'm getting to that!" says Dr. Chris. "Sometimes too many electrons can stick to you. This gives you a negative charge!"

“Uh-oh...” Red Kangaroo says. “Am I going to get an electric shock?”

"Not if the temperature is warm," says Dr. Chris. "The electrons will move off you because warm air has more moisture or water in it. Water is a conductor just like the metal wires hidden in electric cords."

"I thought a conductor controlled train engines," Red Kangaroo says.

"That's a different type of conductor," Dr. Chris says. "This type of **conductor** is anything that lets electric **current** flow easily."

"What happens to the electrons if there is no conductor?" asks Red Kangaroo.

"Then the electrons won't be able to move," Dr. Chris says. "The opposite of a conductor is an **insulator**. Wood, plastic, and rubber are all examples of insulators that stop electrons from moving."

"That explains why I don't get a shock when I pick up an electric cord," says Red Kangaroo. "It's covered in plastic that keeps the electrons safely inside! But I can get a shock if I touch metal?"

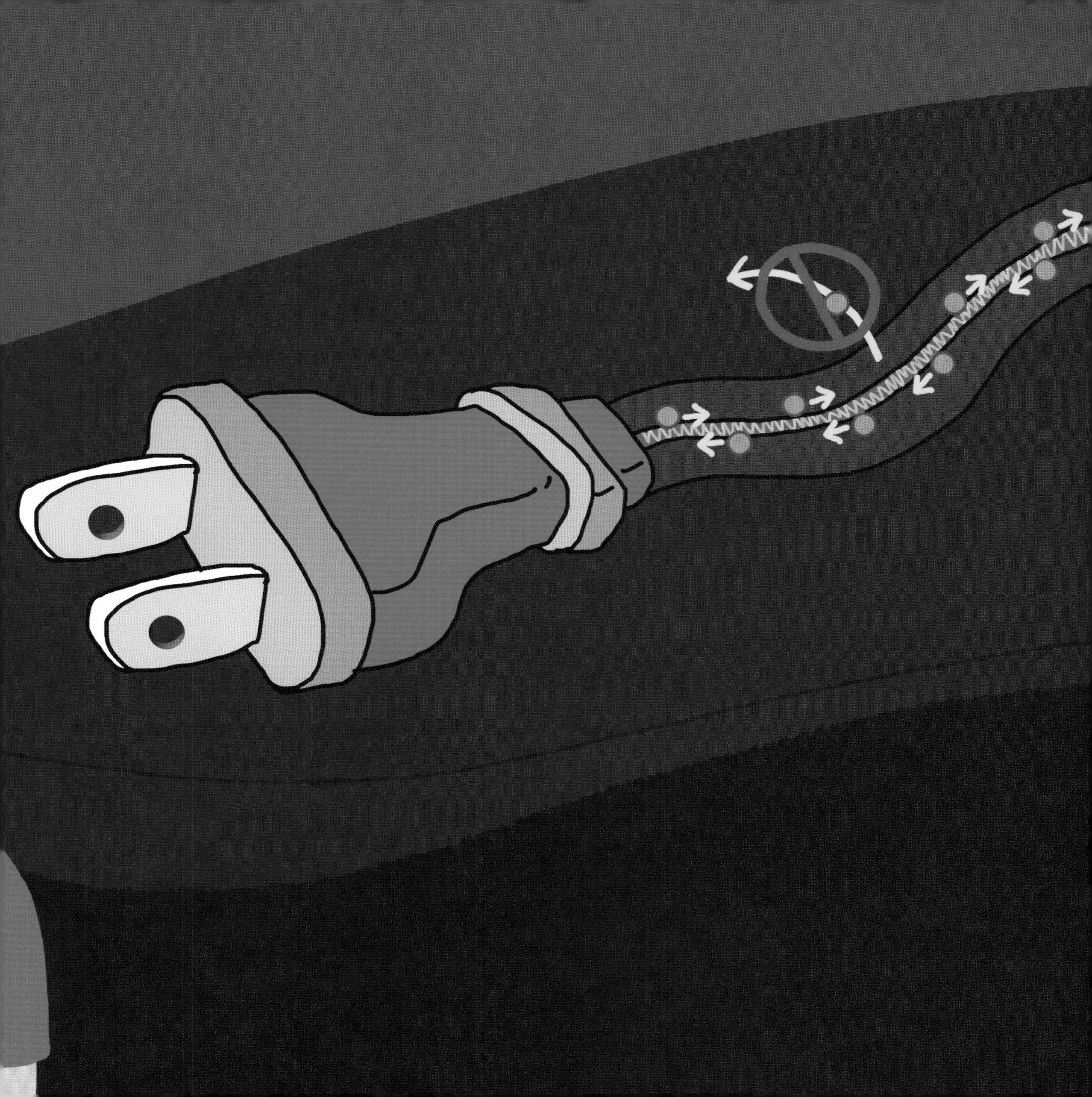

"Yes, but it's not exactly that simple," Dr. Chris says. "An insulator works kind of like filling a cup with water. If I keep pouring and pouring and pouring..."

Dr. Chris

“Oh no!” Red Kangaroo exclaims. “Water spills out!”

"Exactly! An insulator can only hold the electrons until there is too much," says Dr. Chris. "The electrons are free to flow out of the insulator when too many build up. Just like the water did in the cup!"

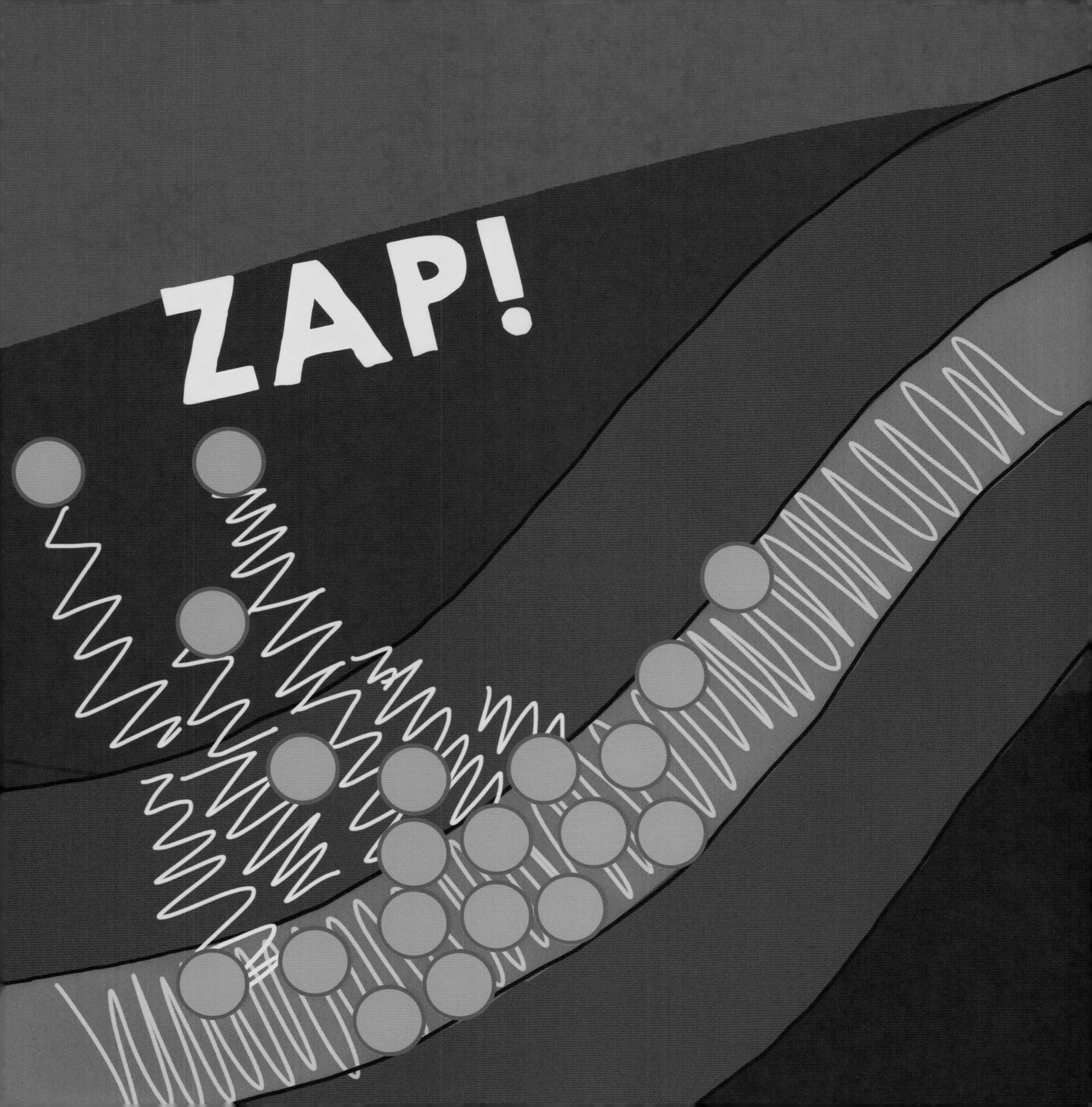
ZAP!

"Cold air is a good insulator because it is so dry," says Dr. Chris. "So the electrons that stick to me build up instead of moving away. And since they have nowhere to go, they keep building and building on my body!"

“Uh-oh. Now your body has a large negative charge, Dr. Chris!” says Red Kangaroo.

"Ouch!" cries Dr Chris.

“It looks like I need to teach you about electric charges, Dr. Chris!” says Red Kangaroo.

Dr.

Glossary

Conductor
A material that allows electric charges to move through it, like copper wire.

Current
The movement of electric charges.

Electricity
A word used to describe all the things affected by charges.

Electron
A fundamental particle that carries one negative charge.

Insulator
A material that prevents charges from moving through it, like plastic.

Proton

A particle that lives in the nucleus of an atom and carries one positive charge.

Shock

A sudden movement of electric charges through a part of your body.

Show What You Know

1. Name the carrier of positive charges. What about negative charges?
2. If you have two charges, how would you know if they were the same or opposite?
3. Why does electricity flow only when something is plugged into an outlet?
4. Are you more likely to get a shock in summer or winter? Why?
5. Can you get a shock if you are already touching a doorknob?

Test It Out

Bending water

1. You will need a plastic straw, a paper towel, and a sink.
2. Turn on the tap so that you get a small but consistent stream of water flowing out.
3. Fold the paper towel in half twice and then around the straw. Stroke the paper towel on the straw ten times. Don't touch the end you stroked!
4. Move the straw towards the stream of water (it shouldn't need to touch the water, just get close to it). Record what happens.
5. Now let the straw touch the stream of water and then move it away and close again. Did the stream of water change from step 4? Why do you think this happened?

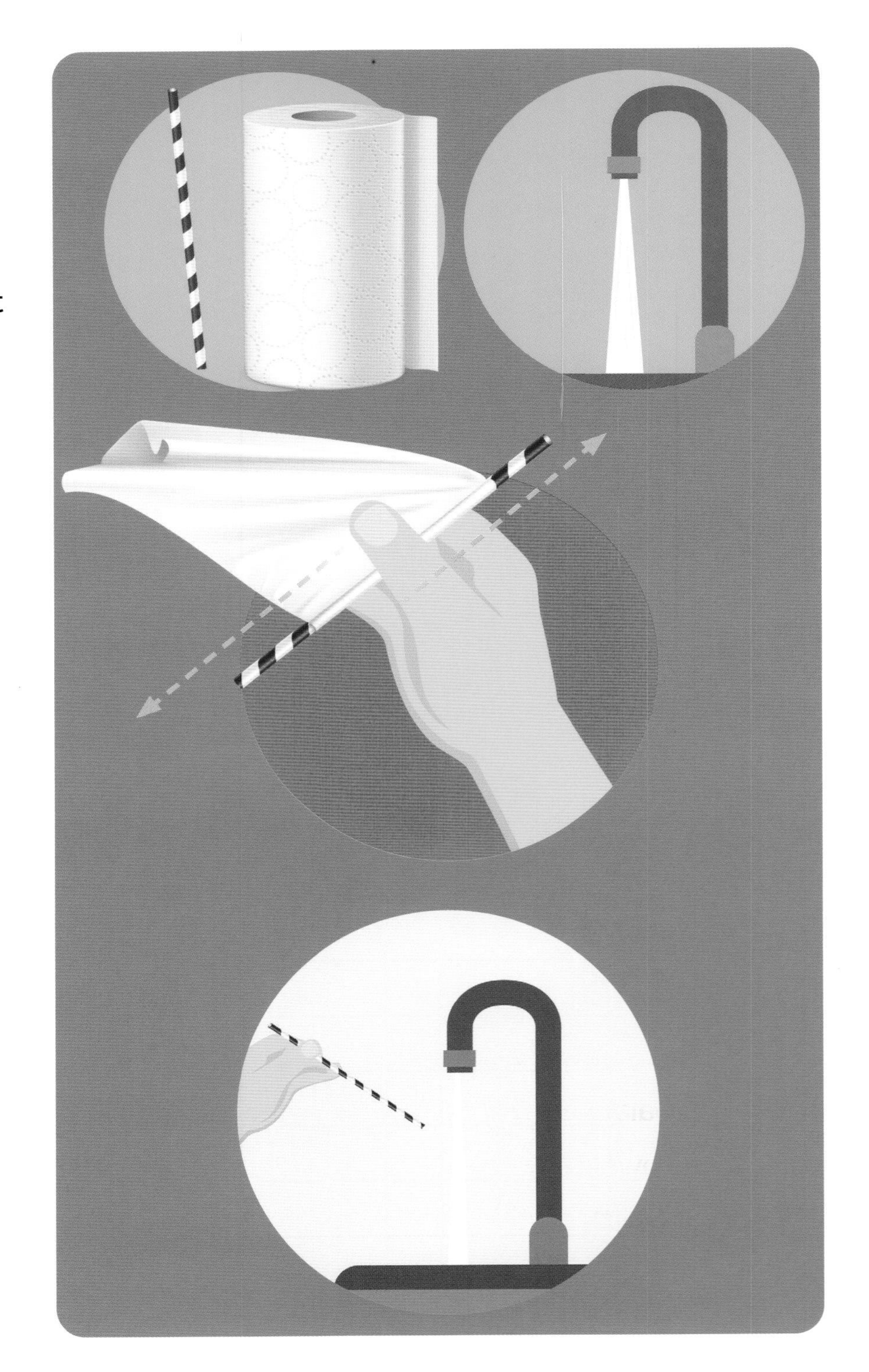

Push and pull of electricity

1. Use the same straw and paper towel. You will need one extra straw and a bottle of water (or something to balance the straw on).

2. Balance the second straw on top of the water bottle.

3. Stroke the paper towel on the straw ten times as before.

4. Slowly bring the straw close to one edge of the straw balancing on the bottle. Watch what happens as you move the straw around the balanced one. Is there a push or pull?

5. Rub the straw that was being balanced with the paper towel ten times and place it back on the bottle. Repeat step 4.

What to expect when you Test It Out

Bending Water

The stream of water should have moved towards the straw. This happens because when you rub objects together (like the paper towel and straw), you get a buildup of electric charge. The rubbing action makes the electrons from one object pass to the other. In this experiment, the paper towel lost electrons and became more positively charged. Water has both positive and negative charges in it. The electrons move to the side of the water near the straw, giving one side of the stream a negative charge and the other a positive charge. This pulls the stream of water toward the positively charged straw. When the straw touches the water, the electrons move onto the straw to return it to a neutral charge.

Push and pull of electricity

This works just like the straw and water. The neutral straw's negative charges are pulled to the positively charged straw you are holding. But! Give them both the same charge by rubbing them and they will repel each other.

Show What You Know answers

1. Protons carry positive charge. Electrons carry negative charge.
2. Opposites attract. If the charges attract, they are opposite. If they repel, they are the same.
3. Current needs a loop to flow. The wire connects the outlet holes to complete the loop.
4. Winter because the air is dry. Dry air has less water, which is a conductor.
5. No. No charge will build up on you if you are touching a conductor to take it away.